Please count with me.

"**Pencils Count**" Copyright 2025

Publisher: Pollywoguen Creations

All rights reserved by Robert J. Carr

ISBN 977-1-959707-26-4 Softback

Produced by Ingram Sparks

Website:

robertcarrpollywoguencreations.com

Instagram: Pollywoguen Creations

PENCILS COUNT

Created by Robert J Carr

1

One

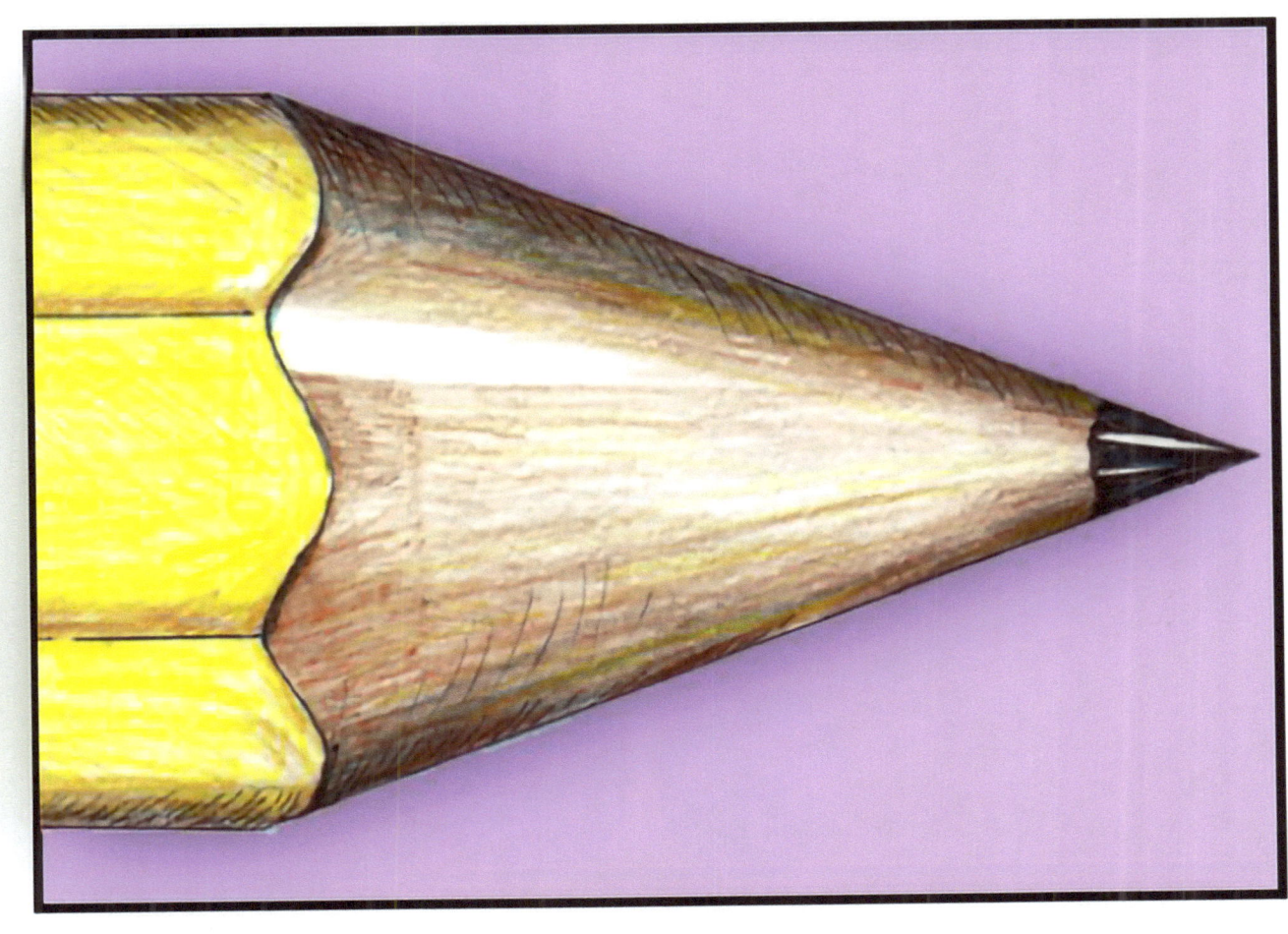

2

Two

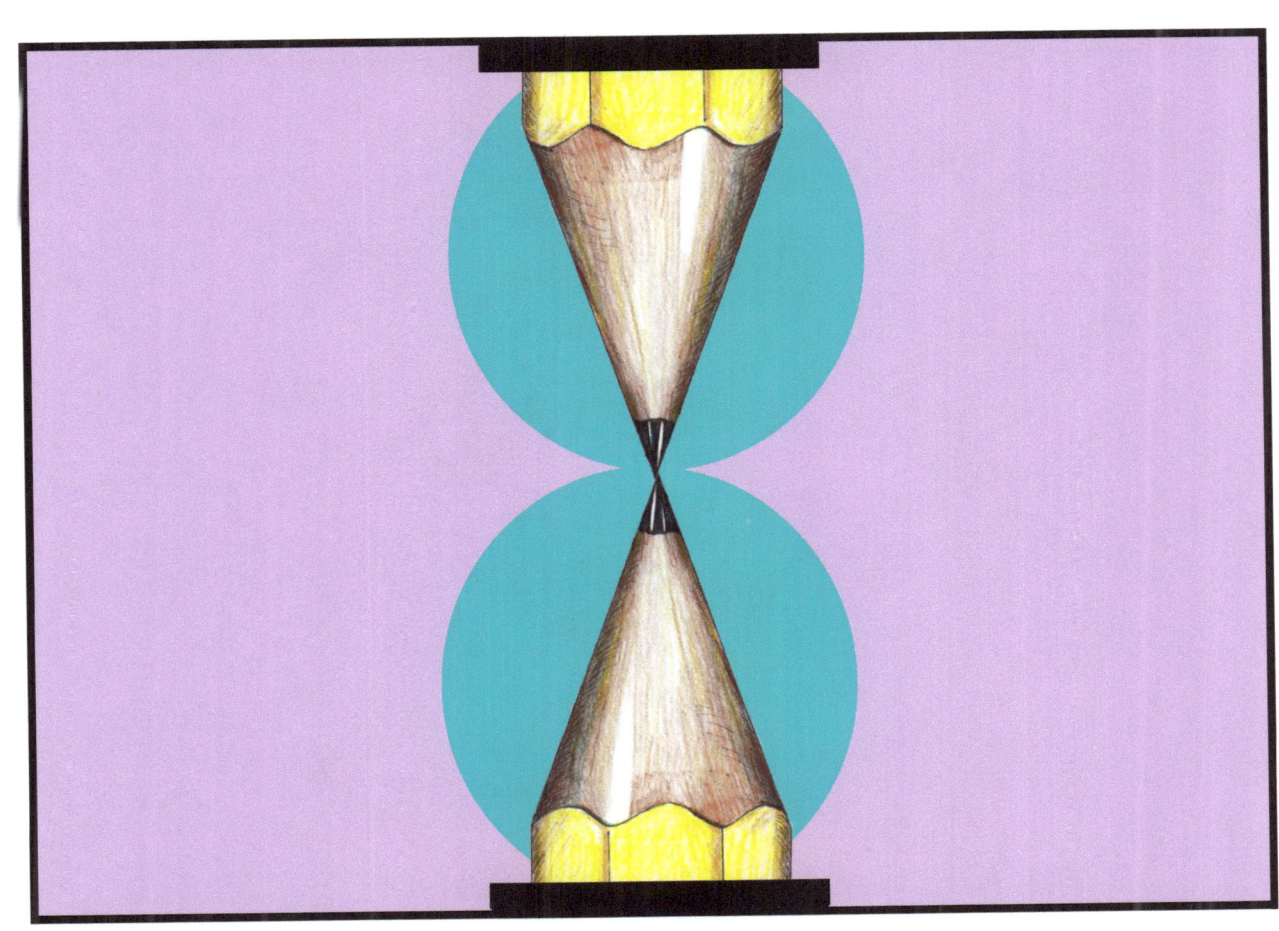

3

Three

Four

●●●●

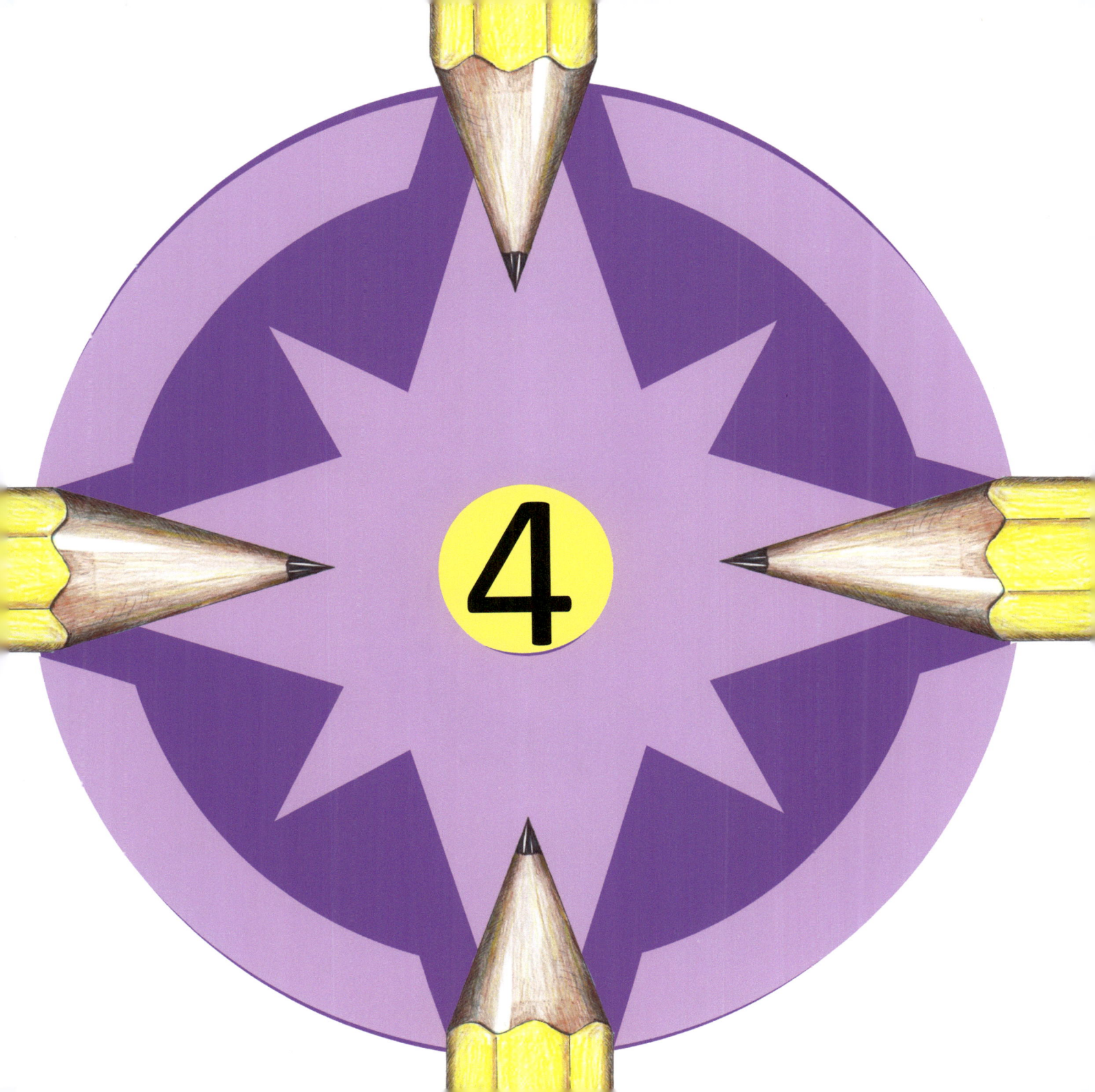

5

Five

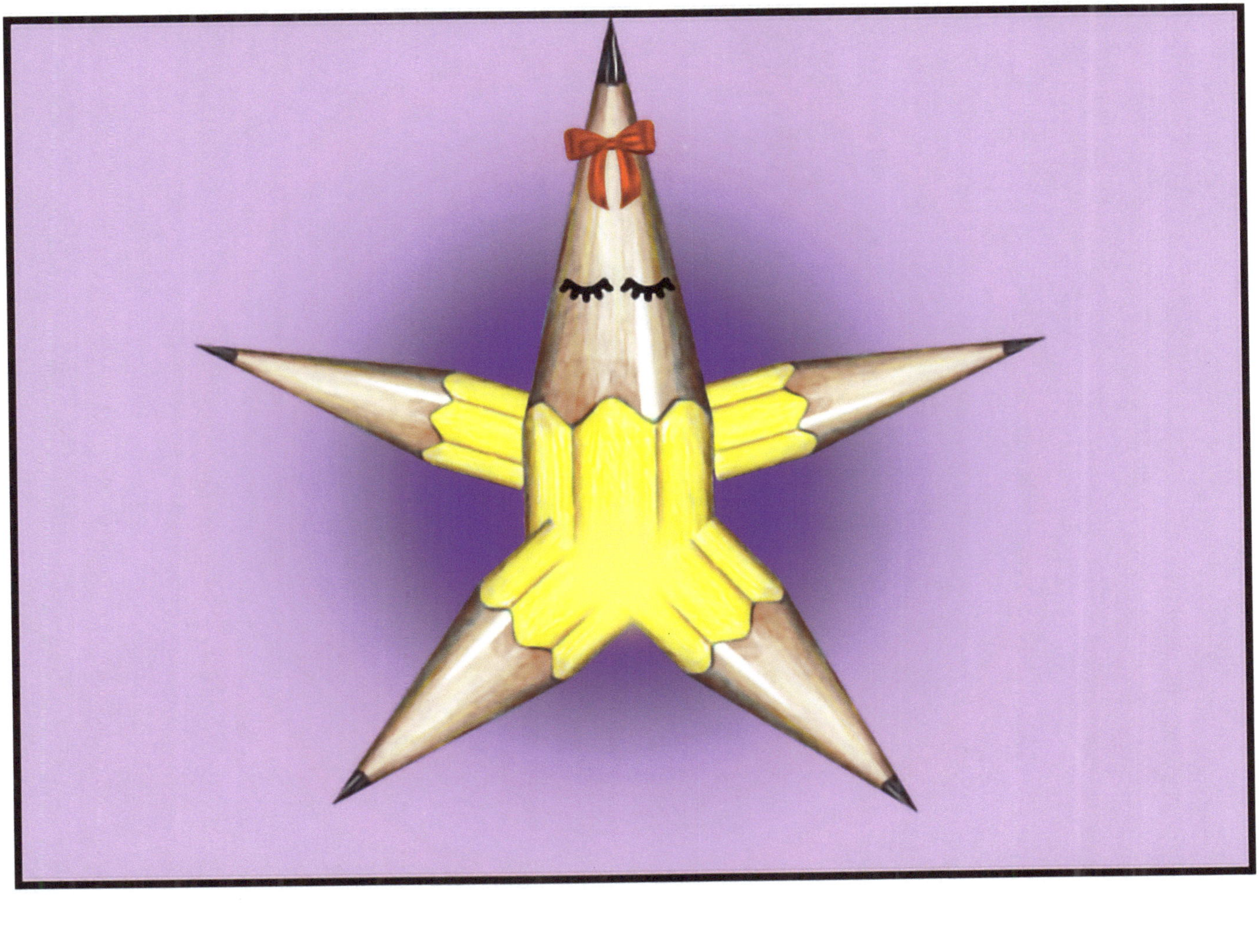

6

Six

Seven

Eight

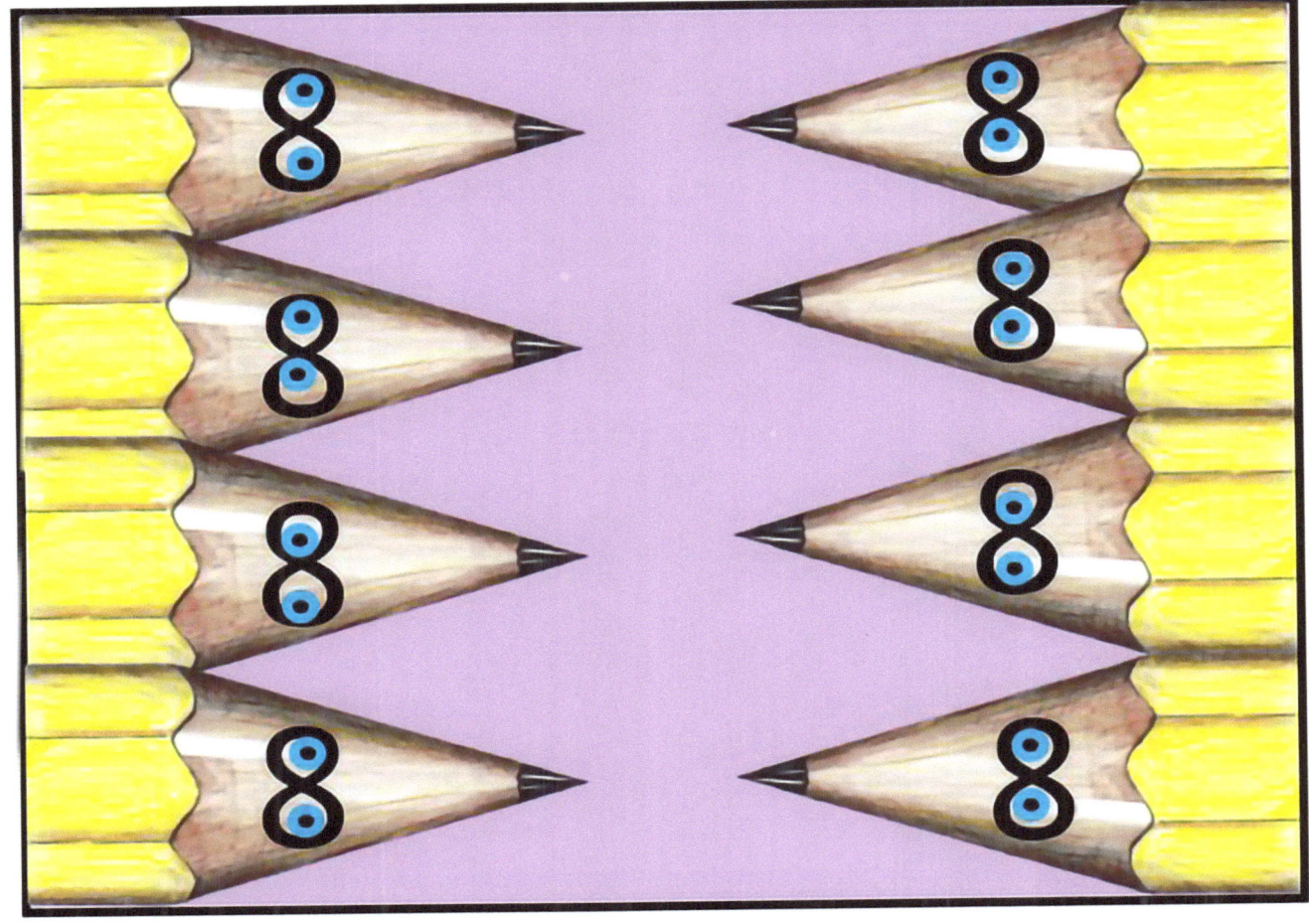

Nine

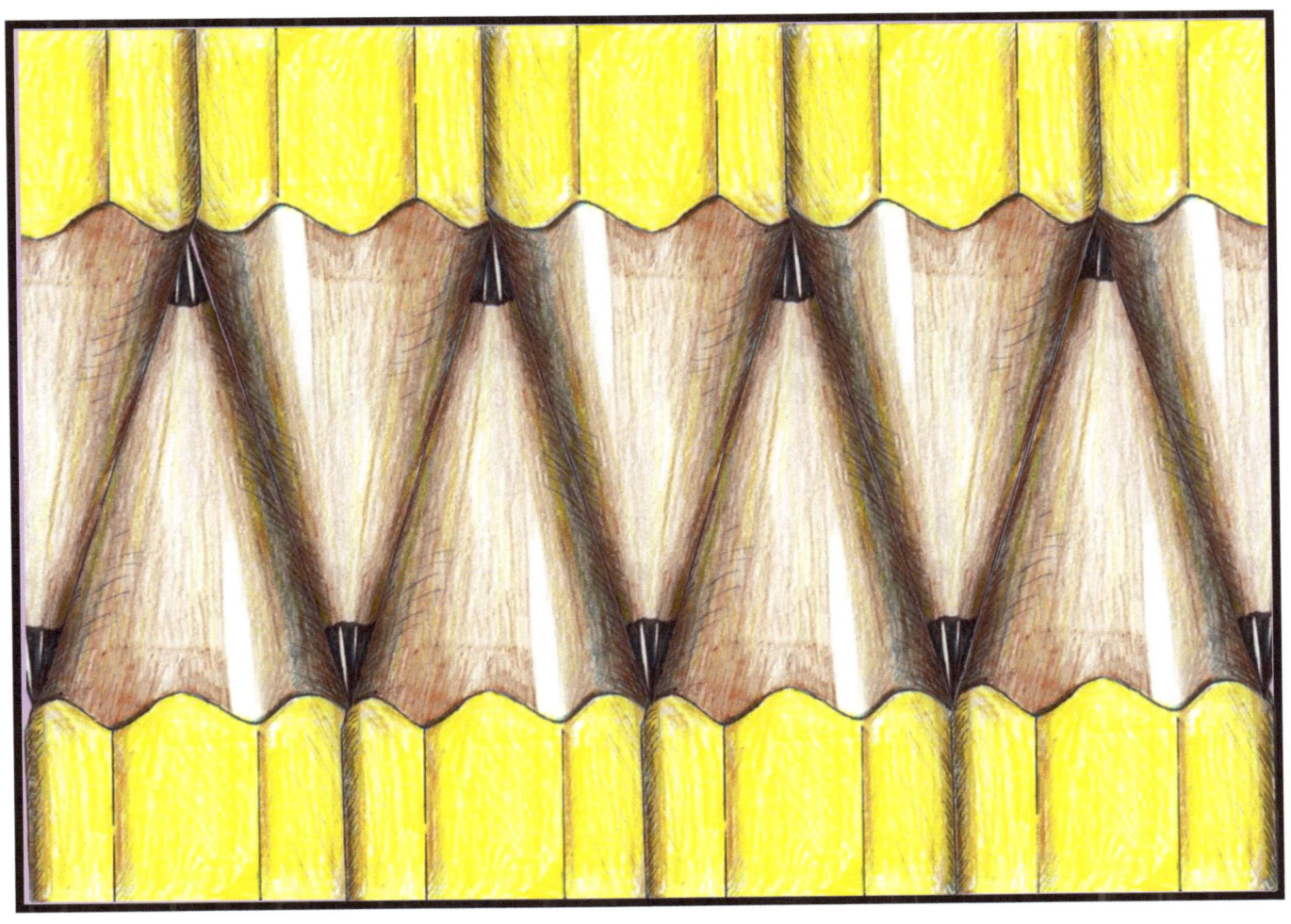

10

Ten

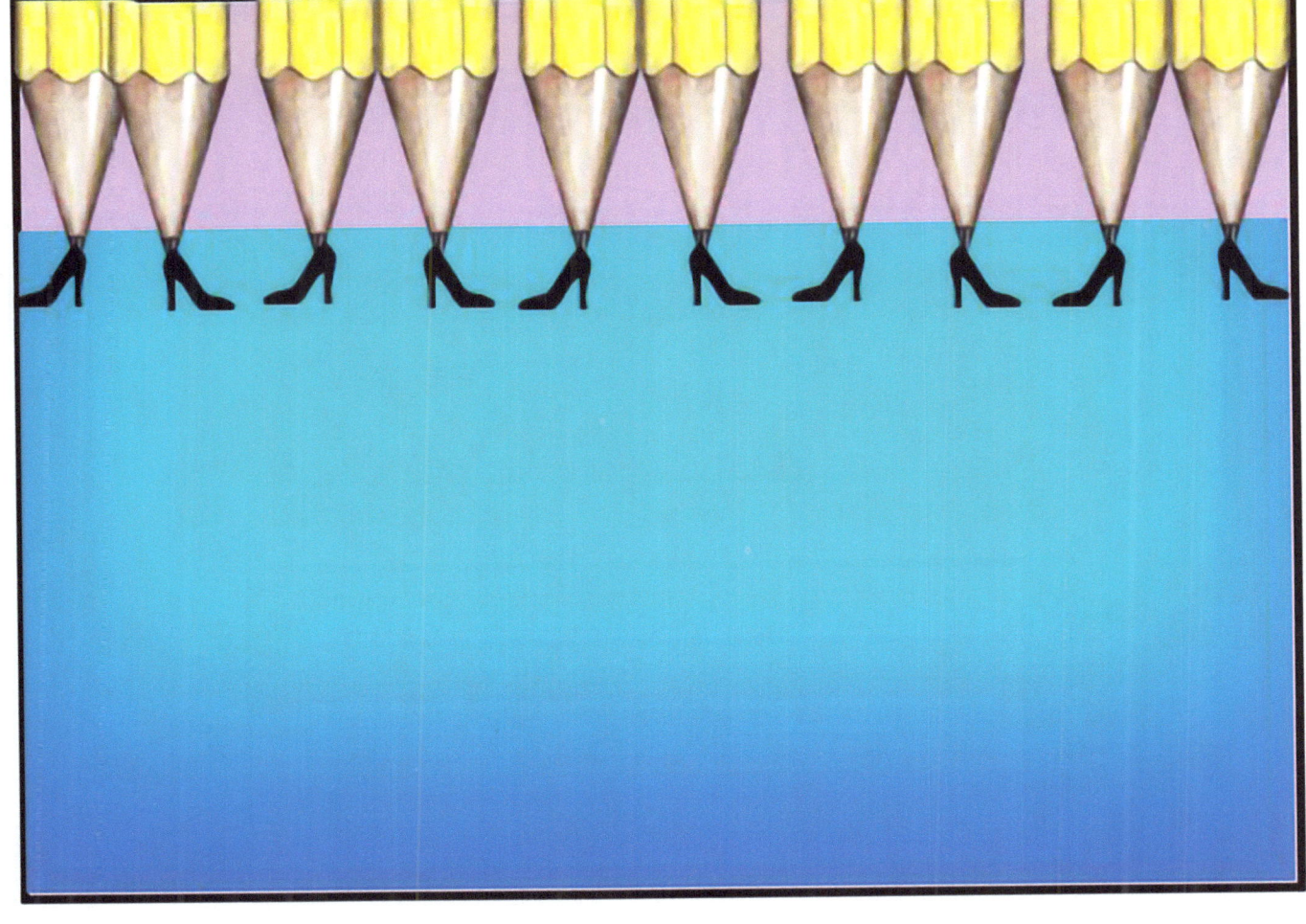

Count, count, count to ten
one to ten and back again
Pictures, words, numbers and dots,
all are here to help you count
May you always have
the right amount
Pencils count and so can you
One to ten and back again.

1, 2, 3, 4, 5, 6, 7, 8, 9, 10

One, Two, Three, Four, Five, Six, Seven, Eight, Nine, Ten

You can count from one to ten over and over and back again. What a gift you now have. Congratulations!

Great Job!

1 2 3

Who on Earth am I?

Other Books by Robert Carr

I Lost My Tooth

SAINT NICK-TIONARY

All About Rainbows

Where is Cleo?

Walrus, Will You Hang With Me?

Hush Little Cyclops Don't You Cry

I am

Uno's World

Pencil Talk

Turkey tales of Jack and Jill Ted and Bill

KELLOGG'S CALENDAR

www.ingramcontent.com/pod-product-compliance
Lightning Source LLC
Chambersburg PA
CBHW041441120626
46547CB00002B/295